GO WILD!
OCEAN EXPLORER

SEAHORSES
IN THE WILD

by Helen Foster James

Cherry Lake Press

Published in the United States of America by
Cherry Lake Publishing Group
Ann Arbor, Michigan
www.cherrylakepublishing.com

Reading Adviser: Beth Walker Gambro, MS Ed., Reading Consultant, Yorkville, IL
Content Adviser: Rachel Morrison, PhD, Animal Behavior Scientist

Photo Credits: © mc_pongsatorn/Shutterstock, cover, title page; © Moonnoon/Shutterstock, cover, title page, 21; © DP0305/Shutterstock, 4; © Kristina Vackova/Shutterstock Images, 6; © Drew McArthur/Shutterstock, 8; © James van den Broek/Shutterstock, 10; © Ethan Daniels/Shutterstock, 12; © patchtheocean/Shutterstock, 14; © Cade Foster/Shutterstock, 16; Tony Wu/Nature Picture Library, 18; © Mariusz Niedzwiedzki/Shutterstock, 20

Cherry Lake Press is an imprint of Cherry Lake Publishing Group.

Cherry Lake Publishing Group would like to acknowledge the work of the Partnership for 21st Century Learning, a Network of Battelle for Kids. Please visit Battelle for Kids online for more information.

Library of Congress Cataloging-in-Publication Data has been filed and is available at catalog.loc.gov.

Printed in the United States of America

Note from Publisher: Websites change regularly, and their future contents are outside of our control. Supervise children when conducting any recommended online searches for extended learning opportunities.

TABLE OF CONTENTS

Seahorses 5

Food and Habitat 11

Families 17

Find Out More.......................... 22

Glossary................................ 23

Index 24

About the Author 24

SEAHORSES

A seahorse is not a horse. It is a fish! Seahorses use **gills** to breathe. **Fins** help them swim. One fin is on their back. Two small fins are on the back of the head.

Seahorses have an unusual shape. They have fleshy bumps that form a **coronet** on their head. No 2 coronets are the same.

THINK

Seahorses swim upright and are slower swimmers than other fish. How do you think they stay safe from faster creatures?

Seahorse eyes can look forward and backward at the same time. Seahorses can also grab with their tail. They use their tails to stay in place.

FOOD AND HABITAT

Seahorses have a long **snout**. They use it like a straw. They suck in tiny sea life that floats by, such as brine shrimp. They can eat 30 to 50 times a day!

Seahorses can change color. This is how they **camouflage** themselves. This keeps them safe.

CONNECT

What other animals do you know that can change color?

Seahorses live in shallow water. They live in seagrass beds. They live in coral reefs. Some live in **tropical** waters. Some can live in colder waters.

FAMILIES

Some seahorse pairs stay together for life. They greet each other every day. They dance together to say good morning!

Seahorse dads carry their young. They give birth to babies. Some can have thousands of **fry** at one time. Others have dozens.

LOOK

How much smaller are baby seahorses than the adult?

Fry are born fully formed. They are small versions of adults. They swim away. In 6 to 12 months, they will find their own **mate**. They will have their own fry.

FIND OUT MORE

Go online with an adult to find videos of seahorses. Write down some things you **see** and **hear**.

Then, write what you **think** about those things. **Make a connection** to something else you know.

Next, write what you **wonder** about seahorses from the videos. What would you like to learn more about?

GLOSSARY

camouflage (KAM-uh-flahj) to blend in with surroundings

coronet (kor-uh-NET) type of crown

fins (FINZ) flaps that are used for moving and steering through the water

fry (FRYE) baby seahorses

gills (GILZ) organs near a fish's mouth used for breathing

mate (MAYT) partner in creating young

snout (SNOWT) the long front part of an animal's head, which often includes the nose, mouth, and jaws

tropical (TRAH-puh-kuhl) related to areas that don't freeze but are warm all year

INDEX

babies, 18–21
behavior, 9, 11, 16–17, 18–19, 21
body parts, 4–9, 10–11

camouflage, 12–13
coronets, 6, 7, 18

defenses, 9, 12–13

eyes, 8, 9

families, 16–21
fins, 5

food, 11
fry, 18–21

gills, 5

habitats, 12, 14–15

reproduction, 18–21

senses, 9
size, 18, 19, 21
snouts, 10, 11
swimming, 5, 7, 14

tails, 8, 9, 16

ABOUT THE AUTHOR

Helen Foster James likes to read, write, travel, and hike in the mountains with her friends. She visited a sea horse farm where a sea horse wrapped its tail around her finger. She lives by the Pacific Ocean in San Diego, California, with her husband, Bob.